Copper Enameling

Another book by LOUIE S. TAYLOR

Successful Soldering

COPPER ENAMELING

by Louie S. Taylor

SOUTH BRUNSWICK AND NEW YORK: A.S. BARNES AND COMPANY
LONDON: THOMAS YOSELOFF LTD

©1977 by A. S. Barnes and Co., Inc.

A. S. Barnes and Co., Inc.
Cranbury, New Jersey 08512

Thomas Yoseloff Ltd
Magdalen House
136-148 Tooley Street
London SE1 2TT, England

Library of Congress Cataloging in Publication Data

Taylor, Louie S
          Copper enameling.

          Includes index.
          1. Enamel and enameling. I. Title.
NK5000.T38   738.4   75-38448
ISBN 0-498-01865-2

PRINTED IN THE UNITED STATES OF AMERICA

This book
is gratefully dedicated
to
Marie, my wife

Contents

Introduction   9

 1 Materials, Tools, and Equipment   13
 2 Characteristics of Enamels   28
 3 Fundamentals of Enameling   31
 4 Counter Enameling   35
 5 Swirling   38
 6 Crackle Enamel   42
 7 Sgraffito   47
 8 Controlled Design   49
 9 Stenciling and Masking   53
10 Restoration   58
11 Foil Overlay   60
12 Cloisonné   64
13 Findings, Soldering, and Cementing   66
14 Silver-Plated Steel   68
15 Thread Drawing   70
16 Glass Beads   76
17 Decorative Gems   85

Index   94

88705

Introduction

Enameling on metal is one of the older forms of handicraft, dating back to the sixth century B.C. Today it is an appealing, leisure-time activity offering an opportunity for self-expression and imagination.

In this book this author makes no attempt to dwell on the traditional techniques or the historical aspects of enameling. As interesting as these phases may be, there is no need to present material already covered in many other books. Instead, this book is concerned primarily with the innovations and recent introductions to the craft.

With these thoughts in mind, this book is divided into a series of chapters depicting the more interesting and gratifying ways of enameling. It is written in a simple, step-by-step, easily understood manner, each chapter relating to one specific feature of the work.

Copper Enameling

1
Materials, Tools, and Equipment

BASE METALS

Copper, gold, silver, silver-plated steel, and aluminum can be successfully enameled.

Copper is the metal most often used by the hobbyist as well as by the professional enamelist. It is easy to use, relatively inexpensive, and can be purchased in a variety of precut shapes. Eighteen-gauge is recommended, either precut or in sheets from which the various shapes can be cut. Unless otherwise indicated, the information on the

An assortment of copper shapes.

following pages will be concerned primarily with copper as a base metal.

Gold and silver are excellent metals for enameling but are expensive and probably should be used only by the experienced enamelist.

Silver-plated steel is a very effective base metal, and the student will enjoy working with it. Chapter 14 will be devoted entirely to a discussion of silver-plated steel enameling.

Aluminum is not used extensively by the enamelist. The melting point of aluminum is lower than that of the regular enamels, and, consequently, special low-firing enamels are required.

Brass and other alloys containing zinc are not suitable for enameling because the zinc component tends to volatilize, producing blisters and bubbles in the enamel.

ENAMELS

Enamels can be purchased at most hobby shops and can be obtained in one or two-ounce packages or in larger quantities. The price is approximately fifty cents per ounce in small quantities and four dollars per pound in larger quantities, certain colors being more expensive than others.

A starter set should contain about ten or twelve colors, predominantly opaques with a few transparents. Each of the colors should be put into a 2-ounce widemouthed jar. A double thickness of nylon hosiery should be stretched over the opening and secured with a rubber band. The nylon serves as a sifter when the enamel is dusted from the jar to the copper, and provides a convenient sieve that assists in removing any contaminating particles when the enamel is returned to the jar.

In order to have a quick and easy identification system each jar should be provided with a sample piece of copper, which has been enameled and glued to the top of the lid. This is best accomplished by using a half inch circle of copper for each of the opaque colors, and a half inch square of copper for each of the transparent colors. The circle can be thought of as signifying an "O" for opaque, and the square, a window for transparent. This procedure provides a good learning experience, acquainting the beginner with the characteristics of each color and with the true appearance of each enamel after it has been properly fired.

ENAMELING KITS

Beginners may prefer to purchase enameling kits. These are available

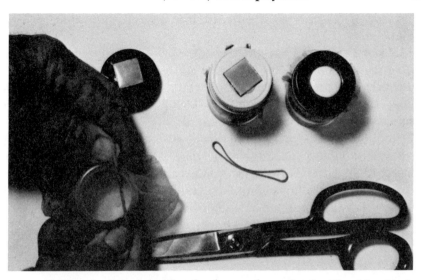

Covering the mouth of the jar with nylon hose and securing it with rubber band.

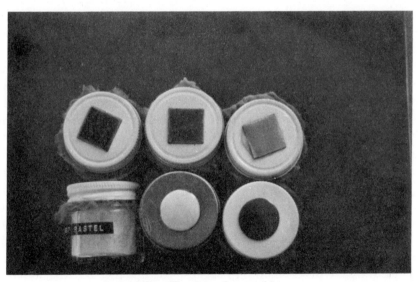

Proper identification of enamel jars.

in craft-supply stores and are reasonably priced but somewhat limited in their usefulness. The electric kiln, which is included, is about five inches in diameter and will serve well for firing most small pieces. The kit usually contains several ready-cut shapes of copper, an inexpensive spatula, about ten viles each holding about a half ounce of enamel, a bottle of 7001 enameling solution, and a swirling wire.

BINDERS

A binder is a solution used to keep the powdered enamel in place when it is being fired. It is applied with a brush or can be sprayed over the surface of the metal. The enamel is then sifted over the binder, causing the enamel to adhere to the surface. Any binder solution that is used must burn out, leaving no residue or discoloration.

There are several solutions that can be used as binders. Gum tragacanth is one of these. It tends to dry very rapidly—a disadvantage in most cases. Klyr-Fire is another type of binder. It has many uses and they will be explained in detail in another chapter. Usually the most satisfactory binder is a solution called "Formula 7001." More information concerning it and other binders will be covered in the following chapters as the need arises.

PICKLING MATERIALS

Any exposed metal surface that is not sufficiently protected or covered with enamel will oxidize when fired and will produce an unsightly fire-scale. This incrustation can be removed by placing the copper in a pickling solution. Sometimes diluted nitric or sulphuric acid is used for this purpose, but it is recommended that the student use a commercial product that is sold under the trade name of "Sparex No. 2." It is cleaner, safer, and superior in many respects. It is manufactured in the form of a dry granular-acid compound. Sparex No. 2 may be purchased at the hobby shop and combined with water according to the directions on the can. A pyrex dish serves as an excellent receptacle in which to prepare it.

PROTECTORS

A commercial product called "Scalex" is manufactured for the purpose of preventing fire-scale. It is a thick liquid-clay product that can be painted on the back of the copper piece before any enameling is done. It is dried under a heat lamp or in some warm place, or it can be dried overnight at room temperature. The front or other side is then enameled in the usual way, retaining the dry Scalex coating on the back. When the article is fired, the Scalex becomes brittle scale that flakes off, leaving a clean surface. This process must be repeated each time a new coat of enamel is applied. Care must be exercised that no Scalex particles get into the enamel.

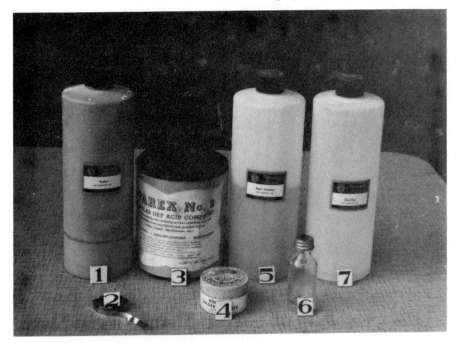

Products used in enameling:
  1) Scalex—used to prevent fire-scale
  2) Ribbon solder for soft soldering
  3) Sparex No. 2—mixed with water it becomes pickling solution
  4) Soldering paste or soldering flux
  5) Agar solution—one of the binders
  6) 7001 enameling solution or binder
  7) Klyr-Fire—another binder

TOOLS

There are a few small tools that are essential in order to do enameling with ease and success. Many of these tools will be found in the average home workshop or can be made by the craftsman or purchased at little expense. The following list will serve to acquaint the beginner with those that are most needed. The tools shown in the accompanying photograph are numbered to correspond to those in the list.

(1)   Tin snips or metal shears
(2)   Ball-pein hammer
(3)   Fret or jewelers saw
(4)   Flat-nose pliers
(5)   Nippers
(6)   Round-nose pliers

(7) Copper-pointed swirling rod
(8) Stainless steel spatula
(9) Stainless steel swirling rod
(10) Palette knife
(11) Tweezers
(12) Small scissors
(13) Copper spatula
(14) Rawhide mallet
(15) Small mill file

EQUIPMENT

Kilns

There are many kinds of kilns that are available to the enamelist. The amateur will probably be interested in obtaining one of the less-expensive varieties, the kind that uses regular house current. The kilns shown in the photographs are all suitable to firing the average piece of enamel, and they work on 115 volts and plug into the ordinary outlet. Some of the larger and more expensive kilns are provided with a pyrometer to

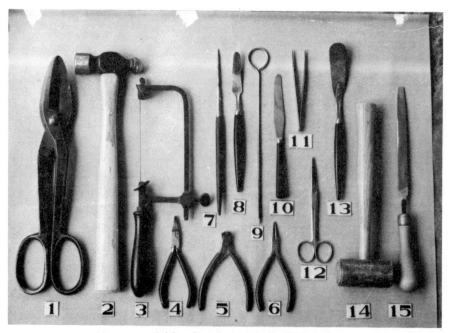

Most essential hand tools as described in the text.

regulate the temperature and to indicate when the proper heat has been reached. The price of kilns range from approximately one hundred dollars to as little as ten dollars. The elements in most kilns are replaceable and, when they burn out or become damaged they can be taken out and new ones installed.

Sometimes after long usage, enamel accumulates on the hearth of the kiln and, when it is heated, the enamel becomes sticky. This condition can be rectified by coating the element with a material known as "Kiln-wash." It is a dry clay powder obtainable in most hobby shops. By mixing it with water a thick paste can be made. The paste is brushed over the hearth to a smooth even coat. It will dry in about an hour and the kiln will be ready for use. This application is a permanent coating that covers the troublesome sticky places and refinishes the surface.

Trivets and Stilts

A trivet is a device that is used to support an object when it is being fired in the kiln. When both sides of the work are enameled, it is necessary to use a trivet in order to prevent the enamel from touching the hearth of the kiln. The piece to be fired is placed in such a manner that only the edge comes in direct contact with the trivet, or it is placed on top of the trivet. If the latter method is used, the enamel on the bottom of the work will stick to the points of the trivet, causing small indentations. These indentations are not usually objectionable and they can be removed easily, or hidden by being smoothed down with a file or covered with small felt pads.

A small electric swirling kiln.

Another type of electric kiln.

Still another type of electric kiln.

A nine-inch by nine-inch electric kiln showing connection to regular 115 volt house current.

Trivets are made in many styles and are usually constructed of stainless steel, a metal that does not scale off when heated.

Stilts are similar to trivets except that they are commonly made of ceramic material and, as a rule, are designed with four steel points. They are obtainable in many sizes, and a glance through a suppliers catalog will give the student additional information concerning them.

Firing Rack

A firing rack made of heat-resistant nichrome wire is used with the larger kilns.

The piece to be fired is set on the rack and the rack is then placed in the kiln.

Firing Fork

A firing fork is necessary in order to transfer the rack to and from the kiln.

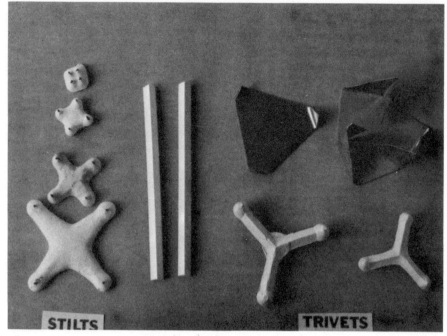

Samples of stilts and trivets.

Torches

A torch is an essential piece of equipment. There are several kinds of torches, but the type that burns acetylene gas from the tank and oxygen from the air will be the most satisfactory.

The bottled propane torch is good for small pieces. It should be held firmly in the hand or secured in some way to prevent it from rolling or becoming uncontrollable.

A special torch used by glass blowers is a convenient piece of equipment and serves well on special occasions. It uses natural gas and compressed air. When adjusted properly, the flame can be applied directly to the enameled surface without discoloring or staining the enamel, this being especially advantageous when transparent enamels are used. This equipment is more sophisticated and less likely to be available. It requires an air compressor and natural gas piped to the location where it is to be used.

Metal Grid

A special grid will be necessary for firing enameled pieces with a

Large spatula and lifting fork used to place enameled pieces in and out of the kiln. Also, two types of nichrome wire firing racks.

The acetylene torch and a "B" tank of acetylene with regulator gauges.

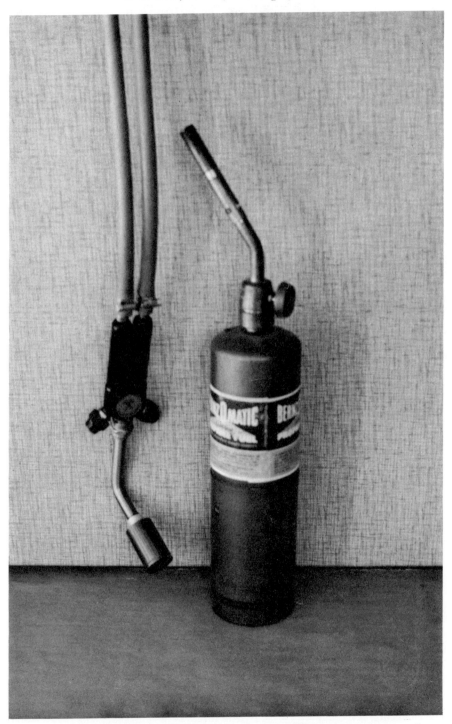

A natural gas and compressed air torch. Also the propane gas torch.

torch. The student can have it made at a welding shop or may be able to construct it himself. It consists of a framework of one-quarter-inch or three-eighths-inch steel rod covered with black metal lath. The metal lath is stretched over the frame and held securely in place by overlapping the edges about one inch.

To make the frame, a piece of one-quarter-inch or three-eighths-inch round steel rod thirty-six inches long is required. Thirteen and one-half inches from each end, the rod is heated and bent at right angles, leaving a nine-inch section of rod between the two corners. Another nine-inch piece of rod is welded between the two thirteen and one-half-inch members forming a nine-inch square with two four and one-half-inch legs projecting beyond. When the framework is completed, the metal lath is turned over the edges forming the grid. The work to be fired is placed on the grid in a convenient way and fired from the underside.

Before it is used, it will be necessary to drill two holes into the bench at a convenient location to accommodate the two legs. This leaves the grid projecting outward from the workbench in an excellent position for firing with a torch.

The enameling grid along with an acetylene torch, a "B" acetylene tank, and a water receptacle.

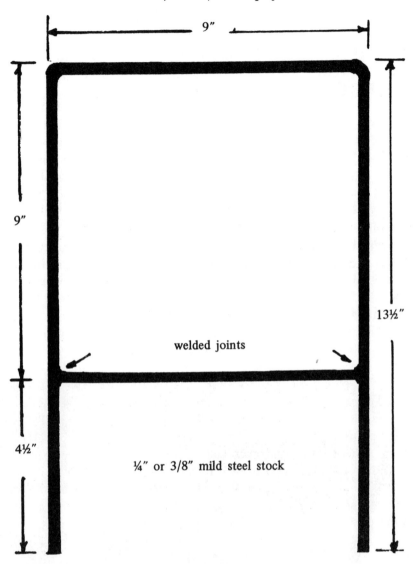

9"

9"

13½"

welded joints

4½"

¼" or 3/8" mild steel stock

Framework for grid.

2

Characteristics of Enamels

The enamel that is used by the enamelist is primarily glass, technically called *silica*. To provide special properties such as luster, fusability, opacity, and elasticity, small quantities of potash (Potassium Carbonate), borax (Sodium Tetraborate), and soda (Sodium Carbonate) are added.

Each individual color is created by the addition of a certain oxide of metal. The reds and oranges are produced by adding gold oxide, accounting for the fact that they are more expensive than the other colors. The green and turquoise colors are produced by adding oxide of copper. The blues, for the most part, are the result of the addition of oxide of cobalt; yellow, uranium oxide; black, iridium oxide; white, tin oxide; purple, manganese oxide; and the grays are obtained by using oxide of platinum.

When a batch of molten enamel is removed from the *retort,* or furnace, it is cooled and broken into small pieces. This raw material is called "frit" and is made in approximately two hundred colors. Frit is ground to various grades of powder or left in small fragments referred to as lumps.

KINDS OF ENAMELS

Fluxes

Enamels that contain no metalic oxides are colorless transparent enamels known as fluxes. There are several kinds of fluxes, each melting at a different temperature and each serving a different purpose. Clear flux when used over any metal will expose the color of the metal. When clear flux is fired over copper, the copper will show through the flux. Frequently flux is used under transparent enamels to enhance their colors.

Each base metal requires its own special flux. There is one manufactured especially for copper, one for silver, another for silver-plated steel, etc.

Opalescent Enamels

Opalescent enamels, sometimes referred to as *translucent* enamels, are semitransparent. The choice of color is limited, but when a compromise between a transparent and an opaque is desired they often serve a definite purpose.

Crackle Enamels

Crackle enamels are sometimes called *slush*. They produce a crackled effect when fired over a prefired undercoat. More information concerning crackle enamel will be found in another chapter.

Painting Colors

Still other enamels are known as painting colors. They consist of a limited number of finely ground powders that are mixed with a special oil and used as paint. A small quantity of powder and oil is placed on a glass slab and, with a palette knife, they are blended to a smooth consistency. Intermediate shades can be obtained by mixing two or more colors. This feature applies to painting colors only.

Opaque and Transparent Enamels

The enamels most commonly used are opaques and transparents. They are usually ground to eighty-mesh and are obtainable in a variety of colors. As indicated by the name, opaques are solid colors that completely conceal the under surface, whereas transparents permit reflection of the light.

The firing temperature of the individual colors varies from 1450 to 1550 degrees F. The enamels that melt in the lower temperature range are termed soft or low firing. Those in the middle group are termed medium fusing or medium firing, and those in the upper range are referred to as hard or high firing.

Every color has its own special characteristics and peculiarities that must be considered. The opaque reds are usually found to be in the low-firing group, and for that reason the danger of overfiring is greater. If overfired they become dull and dark, sometimes even black.

Most of the opaque greens and some of the blues belong to the low-firing group and are subject to color changes with overfiring. They tend to become transparent and somewhat darker. The yellows and blacks are quite stable, usually maintaining their true colors and generally not subject to the problems of overfiring.

Overfiring often causes the base metal to oxidize and the oxides to penetrate the enamel. This penetration is referred to as "bleeding" and it occurs most frequently when the enamel is not thick enough. For example, if a thin coat of opaque white is applied to a piece of copper and the piece is overfired, the copper will oxidize and an undesirable green stain will appear. This usually can be remedied by an additional coat of white and a subsequent refiring, or it can be prevented in the first place by an application of a heavier coat.

In order to be properly fused, transparent enamels require slightly more heat than opaque enamels. Transparents are less likely to burn at the higher temperatures.

The most brilliant transparent colors are achieved when applied over white or flux.

Some enameled pieces, especially black pieces, lose their luster and become dull when allowed to remain in the pickling solution for too long a time.

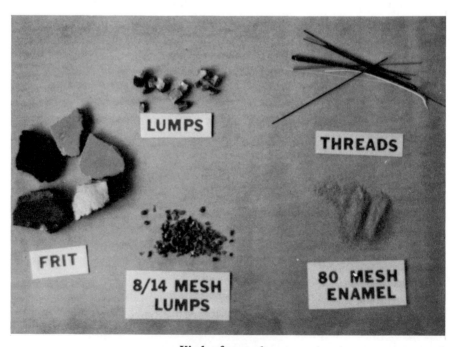

Kinds of enamels.

3

Fundamentals of Enameling

A copper shape is selected. It is cleaned on the upperside with fine steel wool to remove all grease, dirt, and fingerprints. The cleaning is done in an area that is removed from where the enameling is to take place, otherwise small fragments of steel wool may contaminate the enamel.

The copper piece is held in one hand and, with a brush in the other hand, a thin coating of binder is applied. Gum tragacanth may be used, but "Formula 7001" enameling solution is recommended. The student is cautioned against using too much of the solution because it may cause blisters in the enamel during firing.

The piece of copper is placed on a clean, stiff sheet of paper, and a layer of powdered enamel is sprinkled over it. The enamel must completely cover the copper and be distributed evenly. Special care must be taken to dust the enamel out to the edges of the piece of copper.

Insufficient enamel at the edges will result in an unsightly border. It is rather difficult to judge the amount of enamel that is needed. A good rule to follow is to use enough enamel to generously cover the metal, being certain that there are no thinly dusted areas.

The copper piece is now ready for firing, and it can be fired in the kiln or placed on the grid and fired with the torch. A spatula is used to lift the copper from the paper to the kiln or to the grid. The paper is lightly creased between the thumb and fingers to form a slight depression under the copper. The depression provides necessary space in which to insert the spatula and remove the piece.

Firing takes but a few minutes in a preheated kiln. The piece should be watched carefully and removed when the enamel becomes fused and smooth. If the piece is fired with the torch, the enamel is fused by heating it from the underside of the grid.

Metal sculpture that has been enameled.

Handmade glass beads. Hot copper tubing cores are rolled in powdered enamel.

Decorative petal gems placed to represent flowers and fused to background coating.

Controlled design technique. Molten lumps of enamel are shaped by using a copper-tipped rod.

Decorative cattails that have been made by stretching threads over a candle flame.

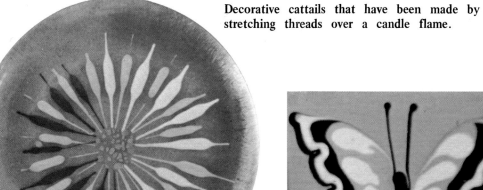

One style of butterfly designed by manipulating molten threads.

Applying **7001** enameling solution.

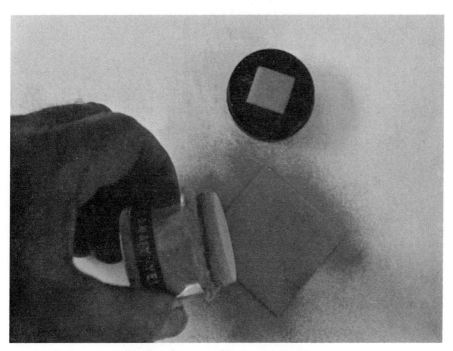

Dusting the enamel on a copper piece.

The work is now placed on an asbestos board to cool. It is then put face-down into the pickling solution where it is allowed to remain three or four minutes. The pickling solution dissolves any fire-scale that may have accumulated on the back of the piece. Excessive pickling should be avoided because it dulls the enameled surface.

The underside of the work should be kept free from particles of enamel as they could be transferred to the kiln. Particles of enamel, which may have become fused to the back of the piece, will not be removed by the pickling solution, but can be removed by using an abrasive cloth or a file.

Enameled piece immersed in the pickling solution.

4

Counter Enameling

Copper expands when heated and contracts when cooled to a greater degree than its enameled surface, creating tension that can cause warping, cracking, or crazing of the article. In order to strengthen the piece and avoid these undesirable features, the piece can be counter enameled on the underside. Counter enamel is dry powder enamel and is available from most enamel suppliers. It is usually gray in color and appears to be a mixture of many colors. Any of the regular enamels can be used as counter enamel and often the enamelist combines his leftover enamels and uses them for counter enamel.

Counter enamel usually is applied to the back of the piece and then fired before any enameling is done on the front. It is necessary to prevent fire-scale from forming on the front of the piece during the firing process. A semiliquid product called "Scalex" is used for this purpose. It is applied with a brush and allowed to dry thoroughly before the counter enameling is done.

COUNTER-ENAMELING PROCEDURES

(1) Both sides of the copper are thoroughly cleaned with "00" steel wool.

(2) Scalex is applied to the face side and allowed to dry.

(3) The back is brushed with 7001 enameling solution and a uniform coating of counter enamel is applied by the sifting method.

(4) The piece is placed in the kiln with the counter enameled side uppermost and fired at a temperature of about 1400 degrees F.

(5) The piece is removed from the kiln and placed on an asbestos board to cool.

(6) After the scalex has flaked off, the face side is cleaned again with steel wool.

Placing counter-enameled piece in the kiln.

Counter-enameled piece cooling on the rack.

(7) The face side is brushed with 7001 solution and dusted with the desired color of enamel.

(8) The piece is centered on a trivet with the counter-enameled side down. The trivet is used for the purpose of preventing the counter enamel from sticking to the hearth of the kiln.

(9) The copper piece with the supporting trivet is placed in the kiln and heated for a few minutes at a temperature of about 1500 degrees F.

(10) When properly fused, the article is removed from the kiln and cooled.

It is not always necessary to counter enamel. In fact, in some instances, counter enameling may be inadvisable. This is especially true when small pieces of jewelry require that findings be soldered to the backs of them. Solder will not adhere to any enameled or counter-enameled surface and can only be successfully used on freshly cleaned metal.

5
Swirling

Swirling, sometimes referred to as *scrolling*, is a technique whereby several different colors of enamel are swirled together to create a random design.

EQUIPMENT NEEDED

(1) The acetylene torch is used for swirling as it provides excellent control of the temperature and facilitates the manipulation of the colors.

(2) A grid is necessary. It provides a place upon which to set the copper piece when it is being fired with the torch.

(3) A swirling rod is needed. The steel swirling rod that is usually furnished with the average enameling kit is not completely satisfactory. It has a tendency to stick to the molten enamel and requires frequent quenching. A better swirling rod can be made by the craftsman by sharpening a two-inch length of twelve-gauge copper wire and hard-soldering it to an eight-inch length of stainless steel welding rod. Copper is an excellent conductor of heat, therefore a rod with a copper tip requires less quenching.

SWIRLING PROCEDURE

(1) The face side is thoroughly cleaned with "00" steel wool.

(2) A thin coat of 7001 enameling solution or other binder is applied to the surface. The binder should completely cover the piece, otherwise the enamel will pull away from the edges on firing.

(3) A layer of enamel is dusted over the piece. The piece can be fired at this time, but it is not necessary to do so.

(4) Small lumps and threads are scattered over the enamel.

(5) With the assistance of the spatula, the piece is placed on the grid.

(6) The torch is lighted and heat is applied to the piece from the underside of the grid. The temperature is slowly increased to a point where the enamel is melted and the surface is smooth.

(7) A random design is created by dipping the rod into the molten enamel and swirling it.

(8) The work is slowly heated to a glowing red. This additional firing levels out the enamel and prevents cracking.

(9) The piece is cooled on an asbestos board and placed in the pickling solution to remove the fire-scale. The edges are filed and the back is polished and lacquered.

Most enamels are darker when they are in the molten stage. Some colors loose their identity and it is difficult to distinguish one color from another. The light and dark colors are more distinguishable and should be used together. Swirling is a random technique and can not be entirely controlled. Over-swirling should be avoided; the final results are more beautiful when the design is simple.

Powdered enamel, threads, and lumps added in preparation for swirling.

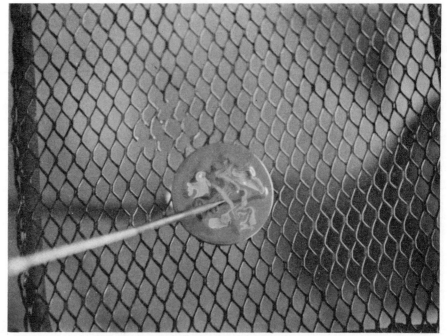

Swirling in progress using a copper-tipped rod.

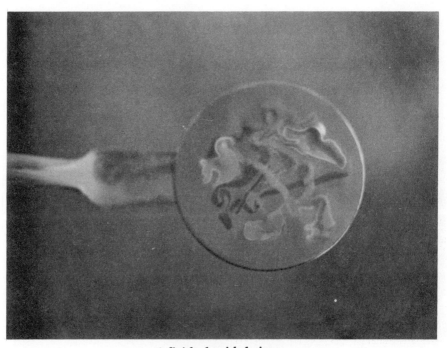

A finished swirled piece.

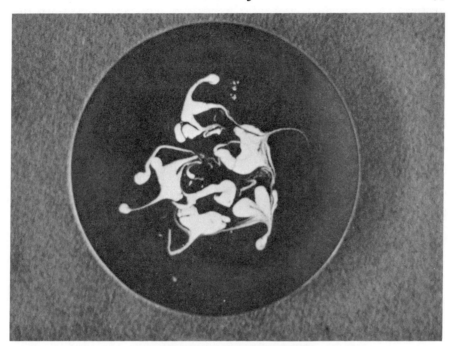

Example of a typical swirl design.

Example of creative swirling.

6
Crackle Enamel

Crackle is composed of finely ground enamel powder, a certain clay product, and water. The main purpose for using crackle is to obtain special, interesting effects over prefired enamel.

Crackle is purchased in semiliquid form and is obtained in small one-ounce jars or in larger jars. The choice of colors is somewhat limited, but there are sufficient colors to satisfy most needs.

Crackle may be applied by brushing, dipping, or spraying. When crackle is fired over regular enamel, the crackle shrinks and pulls apart in places, causing cracks or checks to appear. The undercoat is exposed through these cracks resulting in a contrasting crackled effect. These cracks are random except under certain circumstances, which will be explained later in this chapter.

The piece being crackled need not be counter enameled, but counter enameling will help to strengthen it and provide assurance that it will be less likely to crack at some future time. Cracking or chipping of the enamel most often occurs when the work is enameled on one side only.

Work with crackle should be confined to minor pieces until the student has perfected the technique. Satisfactory results are not obtained when small, flat pieces of copper are used. It is advisable to use domed or convex pieces, which are about one and one-half inch in diameter.

PROCEDURE FOR SIMPLE CRACKLE

(1) A fairly heavy, even coat of crackle is applied over a prefired coat of regular enamel.

(2) The piece is placed aside to dry under a heat lamp or in some other warm place. It must be allowed to dry thoroughly, otherwise

when it is fired the moisture in the crackle will produce steam, which will in turn cause blisters.

(3) The piece is then fired in the kiln at a temperature of about 1500 degrees F., the temperature being maintained until the desired amount of crackle is produced.

(4) The piece is removed from the kiln and allowed to cool on an asbestos board. If the piece is counter enameled it needs no further work. If it has not been counter enameled it is placed in the pickling solution to remove the fire-scale and then cleaned, polished, and lacquered.

PROCEDURE FOR CONTROLLED CRACKLE

The procedure is the same as for simple crackle except that between steps two and three an extra step is added. A design is scratched into the dry crackle, thereby exposing the undercoat. To a degree, the random crackled effect is altered because the crackle has a tendency to follow the lines of the design.

In some cases, removal of large portions of the crackle may be desirable to affect a bold or definite pattern. This can be accomplished with any sharp instrument such as the edge of a palette knife.

The firing and finishing is done as explained under simple crackle.

Applying crackle enamel.

Example of simple crackle.

Example of controlled crackle.

Controlled crackle design.

Additional example of controlled crackle design.

Controlled crackle design.

7
Sgraffito

Sgraffito is an art whereby a second coat of enamel is applied and then partially scratched off, exposing the undercoat and producing a particular pattern or design. Either opaque or transparent enamels are suitable and may be used separately or in combination with each other. There are two popular methods by which this technique may be utilized with good results. The procedure is simple but it is essential that the proper sequence be followed.

The first method should be used when a phantom effect is desired.

(1) The piece of metal is thoroughly cleaned, 7001 solution is applied, and the work is enameled and fired in the usual manner.

(2) When the piece is sufficiently cooled, it is sprinkled with a light coating of enamel that usually is contrasting in color. No oil or binder is used and the piece is not fired.

(3) With the aid of a stylus, a design is sketched or scratched in the enamel, which exposes the undercoat. If the enamel is too thick it tends to accumulate ahead of the stylus, resulting in an irregular edge that may be objectionable. A small sharpened stick of bamboo makes an excellent stylus and is preferable to a metal one.

A piece of cardboard or heavy paper cut to resemble saw teeth can be zigzaged across the work, creating an interesting pattern. A comb can also serve effectively in the same manner.

(4) The work is now ready for firing and finishing.

The second method is used when a more distinct design is needed or one in which larger areas of the base coat are to be exposed.

(1) The copper piece is cleaned, enameled, and fired as directed in step one of the first method.

(2) Klyr-fire or gum tragacanth is sprayed sparingly or brushed over the pre-enameled surface and another coat of enamel is applied.

(3) The piece is allowed to dry until a light crust is formed and the enamel clings to the binder. At normal room temperature this will require from one to two hours, but the time can be shortened to a few minutes by placing the work under a heat lamp.

(4) When sufficiently set, the crust is scratched with a stylus, sketching in the design. The unwanted portion of enamel is removed and the undercoat is exposed in the appropriate places. Extreme care should be exercised to remove all loose particles of enamel.

Sgraffito work using a paper comb.

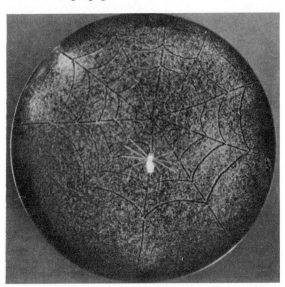

Phantom sgraffito work.

(5) When satisfactory results have been achieved, the work is ready for firing.

If fired in the kiln, the article can be counter enameled before any other work is done. If the firing is done with the torch, the counter enameling must be omitted and the back cleaned, polished, and lacquered.

Sgraffito is a technique that can be used advantageously for making initials, monograms, or scriptwriting.

8
Controlled Design

This aspect of enameling offers an exceptional opportunity for self-expression and imagination. The beginner will wish to explore the many possibilities, but instructions will be given here for one specific example of controlled design. The poinsettia flower employs the typical controlled design technique and is a good illustration of what can be done. The necessary steps used in making the poinsettia are as follows; they merit strict attention if the final results are to be successful:

(1) A one and one-half-inch copper disc or circle is used for this purpose. First the copper is cleaned and a thin coating of 7001 enameling solution is applied. A modest coat of opaque white enamel is carefully dusted over the piece.

(2) Ten to fifteen irregular-shaped, opaque, red lumps about the size of this capital "O" are selected. They are placed in a circle midway between the center and the edge of the copper disc. Care must be exercised to avoid disturbing the undercoat as it has not as yet been fired.

(3) In the third step, the work is placed on the grid and from the underside it is heated with the torch until the lumps are fused and soft enough to be moved about. A swirling rod is then used to pull the outer edge of each individual lump toward the rim of the disc. The torch flame should be used continuously in order to maintain the necessary temperature. A little twist of the rod will assist in shaping the outer half of the petal. When a stainless steel rod is used it should be quenched frequently in cold water to prevent it from sticking to the enamel. Less quenching is needed when a copper-tipped rod is used, the construction of which is explained in chapter five.

(4) The lumps are pulled into the center, making them appear more like poinsettia petals. This procedure produces an accumulation of unwanted enamel.

(5) A stainless steel rod is heated red hot and dipped into the accumulated enamel, causing the enamel to adhere to the rod. The rod is lifted upward and a thread of enamel is drawn. The torch flame is used to separate the thread from the work.

(6) The sixth step provides for the addition of the yellow center particles. A lump of yellow enamel is removed from the assortment, wrapped in a heavy piece of paper, and with a hammer it is broken into small fragments. About a dozen of these tiny fragments are selected and

placed close together in the center of the flower. The work is reheated until all of the enamel is level and smooth.

(7) At this stage the final cleaning and finishing takes place.

Enamel and lumps added in preparation for controlled design poinsettia.

Shaping the petals with a copper-tipped rod.

Drawing the petals to the center.

The finished poinsettia.

Example of controlled design.

Example of controlled design.

Additional example of controlled design.

Example of controlled design application.

9
Stenciling and Masking

Stenciling and masking are closely related and are important phases of the enameling art. They offer distinct possibilities for creativeness. In making a stencil, a design is cut out leaving open spaces and lines through which the powdered enamel is sifted. Masking, on the other hand, is the reverse process; the design is formed by covering certain areas and dusting the enamel between and around them.

In most instances, the stenciled or masked article is fired in the kiln rather than by means of the torch. When the torch is used the article cannot be counter enameled.

If the student wishes to counter enamel he must do so before the stenciling or masking is started. If he chooses to forego counter enameling, he can finish the back of the work by cleaning, polishing, and lacquering it.

THE OPEN STENCIL

A design is drawn or traced on a sheet of tracing paper. A scissors, Xacto knife, or razor blade is used to cut out the stencil. If a knife, or razor blade is used, the work should be placed on a sheet of glass. This will make the cutting easier and there will be less tendency to tear the paper. The upper surface of the stencil is oiled with 7001 solution. The oil serves as a medium for holding the particles of enamel that might otherwise fall into the work when the stencil is being removed.

The face side of the work is enameled with a background coating. It is brushed with oil and the stencil is carefully placed in the proper position on the enameled piece. A selected color of enamel is sifted over the stencil, building up enough powder to satisfactorily cover the open spaces. The stencil is then carefully removed exposing the design under it. At this point the work is ready to be fired.

Sometimes a stencil is needed for a curved or flared piece such as a tray or bowl. It is difficult to make a piece of paper conform to the curved surfaces of a bowl, therefore, slits are cut in the edges of the paper making it possible to overlap the sections. This allows the stencil to take the shape of the bowl. It is pressed into the bowl and the overlaps are taped to secure them. Then it is removed and the stencil is sketched in and cut.

Cutting out stencil with Xacto knife.

SIMPLE MASKING

When simple masking is done, the entire piece is enameled and then portions of it are covered with a masking material. In this way the colors under the masked portion are retained when a second coat of enamel is sifted over the piece. To be applied properly the masking material must lie flat on the enameled surface. This is necessary in order to prevent the powdered enamel from falling under the edges.

Natural objects such as leaves, ferns, and moss can be successfully used for masking. Paper cutouts, either original or traced, are probably used more often because they are easier to handle and can be made to depict any desired motif. Flowers, birds, boats, trees, and animals are among the favorite cutouts, and excellent results can also be obtained by using geometric and freeform designs.

Directions for doing simple masking are as follows:

(1) The copper is cleaned and enameled.

(2) After a thin coat of 7001 solution has been applied to the entire surface, the masking material is placed in position.

(3) The mask is brushed with 7001 solution, careful attention being given to the edges.

(4) The enamel is sifted over the exposed spaces making sure that they are uniformly covered. Any enamel that accidentally falls on the mask is not objectionable, it will cling to the oiled surface.

(5) The mask is carefully removed and the article is ready for firing.

SOPHISTICATED MASKING

When the design includes several areas, each being of a different color, a more elaborate procedure is followed. Each area is treated individually with its appropriate color and separate mask. Finally, when all areas are covered with enamel, the firing can take place.

Let us reproduce the scene depicting a light blue sky, snow-covered mountains, a blue lake, foreground in two or more colors of green and details such as rocks, trees, and clouds. A special technique is necessary and certain steps must be observed to assure results that are satisfactory. They are as follows:

(1) The outside dimensions of the mask are determined by laying the selected piece of copper on a sheet of paper and tracing around it. This sheet of paper will be referred to as the "master sheet."

(2) In the outlined space on the master sheet the scene is sketched. Be sure to include all of the details such as the trees, the rocks, and the snow on the mountains. Each area of the scene is labeled with a number that indicates the order in which it will receive treatment. The foreground will be the first to be covered with powdered enamel so it should be number 1. If there is more than one part to the foreground each part should be numbered. The scene shown here has two parts to the foreground, therefore, the left side is number 1 and the right side is number 2. The lake is number 3, the mountains number 4 and the sky number 5.

(3) Five copies of the master sheet are made on tracing paper, omitting the numbers in the sections and omitting minor details.

(4) These tracings are numbered from 1 to 5, and the numbers are placed in the upper righthand corners of the borders.

(5) Number 1 tracing should have area number 1 or the left portion cut away, leaving in the mask all but this part. Number 2 tracing should have area number 2 or the right foreground portion removed. Number three should have the lake section cut out and removed, number four the mountains, and number five the sky. Each tracing should have a section cut out corresponding to its number on the master sheet.

(6) With the five separate masks satisfactorily made we can now proceed to the enameling. The copper base is cleaned and a coat of

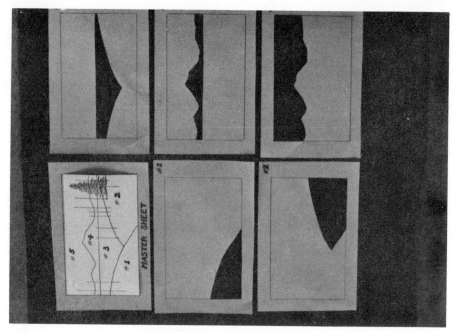

Making a sophisticated mask.

7001 solution is brushed over it and also over the face side of the five masks.

Mask number 1 is placed in position on the copper and held firmly in place with one hand, while the selected green enamel is sifted over the exposed portion of the foreground. The mask is then carefully removed leaving the enamel undisturbed. Next, mask number 2 is gently placed over the copper and, in a similar fashion, the exposed area in it is covered with the other green enamel. The mask is removed and the same procedure is followed with each mask until the entire piece of copper has been covered. Any imperfections in the work can be

Final results with the use of sophisticated mask.

eliminated readily by the application of a slight pressure with the palette knife.

The snow can be added by picking up a small amount of white with the palette knife and placing it on the mountaintops. Small patches of white can be dropped in the sky for clouds. By using threads of a contrasting color the tree trunks can be added.

The work is now ready for the first firing and should be placed in the kiln and fired at about 1500 degrees F. When it is smooth and completely fused it is removed and allowed to cool.

At this time the foliage is added by picking up a small amount of enamel on the edge of the palette knife and gently dusting it in place. Any other fine details are also taken care of. A bit of chartreuse or other light color can be added to highlight the trees, and darker shades can be put in for shadows. With everything completely finished the work is now ready for the second and final firing.

Picture made by using the sophisticated masking method.

10
Restoration

The torch is used in the restoration of work that has been rejected or discarded because of chipped or cracked enamel, displeasing colors, overfiring, or other imperfections. In many instances the copper piece is worth salvaging. If the piece has not been counter enameled and is small enough to be heated with a torch, something usually can be done to restore it.

Let us first consider the piece that has been sparsely enameled and has a coating that is too thin. It may be restored simply by applying additional enamel. A sufficiently heavy coat of any opaque color will completely obscure the undercoat. The addition of a transparent color will let the undercoat show through, sometimes producing a desirable color effect.

When the enamel is too thick, it is possible to remove some or all of the unwanted portion. To do this a stainless steel rod with a sharpened point is used. The work is refired and the point of the rod is heated red hot and dipped into the fused enamel. When the rod is firmly attached to the enamel, a small amount of it is removed by lifting the rod upward and pulling out a thread. The thread is then drawn over to the edge of the piece and is melted off with the flame. More enamel can be removed in the same way, the process being repeated as many times as is necessary.

If the background color is acceptable, but the composition or design is displeasing, the problem may be solved by adding some threads or lumps and swirling them. In this way a fresh and beautiful effect can be achieved.

An overfired piece usually can be restored to its proper color by refiring with an additional coat of the same enamel. Often low-firing enamels that have been overfired can be restored by simply refiring at a lower temperature.

Removing unwanted enamel from copper piece.

When a discarded piece is unacceptable because of cracks or chipped edges it may be repaired. Using a torch, the piece is heated to a bright red and, with a copper spatula, the melted enamel is carefully spread over the damaged places. The piece is then refired in order to level the enamel.

Upon occasion it may be desirable to completely remove the enamel. This may be accomplished by using either the kiln or the torch. The rejected article is heated until it becomes red hot and then is immediately plunged face downward into icewater. This will cause most of the enamel to crack off. It may be necessary to repeat this treatment in order to remove all of the enamel.

There is still another method that may be used in removing unwanted enamel. The unsatisfactory piece is placed on the grid and is heated to a bright red color. This high temperature must be maintained while the enamel is being removed. A stainless steel rod is laid flat on the surface of the piece and is pushed across the piece, gathering up the enamel. When the accumulated enamel reaches the edge, the rod and the adhering enamel are lifted upward and plunged into cold water. The quenching action causes the enamel to crack off.

A thin film of enamel may remain on the work. This is not usually objectionable as it can serve as the initial coat when the piece is redecorated.

11
Foil Overlay

Spectacular effects are obtained when foil overlays are embedded in enameled pieces. The following statements relate to foil overlays.

(1) Gold, silver, and copper foils are most commonly used. Copper is used more extensively than either of the other two because it has the advantages of being thicker, less expensive, and easier to handle.

(2) Copper foil is obtained in hobby shops and is sometimes called *tooling copper*. Thirty-six gauge is ordinarily preferred but lighter metal is also suitable.

(3) Gold and silver foils are seldom found in the ordinary hobby shops. Usually they must be ordered through a catalog. They are manufactured in sheets each approximately four-inches by four-inches and separated by protective papers.

(4) Either opaque or transparent enamel is applied to copper foil overlay. It is recommended that transparent be used on gold or silver as this permits the beautiful tones of the metal to be seen through the enamel.

(5) Tweezers are needed when working with foils. This is especially true in the case of gold and silver foils since they are very fragile.

(6) A small, sharp scissors is needed to cut the foil overlays.

(7) A torch or kiln can be used to fire the piece that is overlayed with foil.

(8) The edges of the overlay may curl up when the piece is being fired. This can be controlled with a steel rod or spatula.

COPPER FOIL OVERLAYS

The foil must be cleaned on both sides with steel wool as both sides make contact with enamel.

In most cases, the design for the overlay is first sketched on a sheet of tracing paper. The paper is then glued to a piece of foil by using white water-soluable glue. When the glue has dried sufficiently the design is cut out and placed in water to dissolve the glue and to permit the removal of the paper.

If so-desired, the design can be cut out freehand from the foil, or it can be sketched directly on the foil and then cut out.

There are two methods used in applying foil overlay:

First method:

(1) The foil is oiled with 7001 solution and covered with powdered enamel.

(2) The foil is placed on the previously enameled surface.

(3) The entire piece including the foil is fired.

Second method:

(1) The foil is oiled and placed on the previously enameled surface.

(2) Enamel powder is then sifted over the foil.

(3) The entire piece including the foil is fired.

(4) When opaque enamel is used, special care should be taken to remove any scattered particles that may fall on the undercoat.

If transparent enamel is used, beautiful effects can be achieved by purposely sifting the enamel over the background as well as over the foil. Clear flux can be used in the same way and the natural color of the polished copper will be revealed.

GOLD AND SILVER FOIL OVERLAYS

Since gold and silver foils are thin and delicate, much care must be exercised to prevent them from folding and wrinkling. Any distortion may show in the finished piece and this should be avoided if possible. The foil is most efficiently handled when it is sandwiched between the two sides of a folded piece of tracing paper. The design is sketched on the outside of the folded paper and the foil is slipped down to the folded edge where it is held securely. This technique protects the foil and keeps it in place while the design is being cut out, and also eliminates the risk of fingerprints.

With the aid of tweezers, the foil is carefully separated from the paper. The overlay is then pricked with a needle in several places. This procedure forms vents for the purpose of allowing trapped air to escape when the piece is fired. Now the foil overlay is picked up with the tweezers and placed in position on the oiled surface of the enameled background. The overlay is also oiled and smoothened.

If flux is used, an interesting effect is achieved by revealing the natural color of the foil. If transparent color is used the gold or silver shows through, but the color will be primarily that of the selected enamel.

Special attention is required when gold and silver overlays are fired because of their comparatively low melting points.

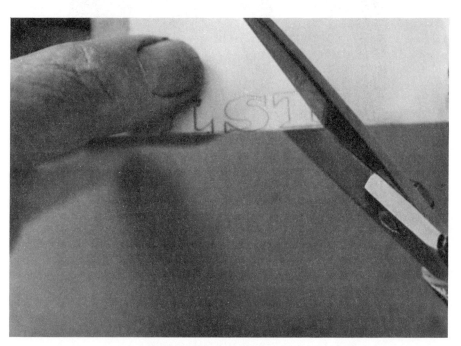

Cutting out letters from silver foil.

Finished piece using silver foil and gems.

Finished piece using copper foil and lumps.

12
Cloisonné

Cloisonné is one of the older forms of enameling. It was done by the Greeks more than two thousand years ago. Today, however, the techniques and methods are different. Enamels are greatly improved and the enameling process has been simplified.

The first step in doing a cloisonné is to select a piece of copper the desired size and shape. It is enameled on one side with a background coat of either transparent or opaque enamel.

One or more pieces of copper wire are cut and shaped to form a planned design. Twenty-gauge copper wire is used and is shaped with the aid of small round-nose pliers. The wire is hammered lightly on a flat steel surface. Hammering hardens and flattens the wire and produces a greater area of contact with the base.

The sections of wire are placed in the correct positions on the enameled piece, each section being held in place by applying a small amount of white glue at strategic points, such as the ends of all pieces and the centers of the longer ones. The glue is used sparingly and is applied with a small sharpened stick of bamboo or a toothpick.

The work is allowed to dry for several hours or until firm, and then it is fired to burn out the glue and to embed partially the wire in the base coat of enamel.

The small enclosures referred to as cloisons are now filled with enamel. Each color is prepared as needed by placing a small amount on a sheet of glass and, with a palette knife, carefully blending in a few drops of klyr-fire. The consistency of this mixture must be just right, not too thick and not too thin, in order to fill properly and neatly the cloisons. After each section has been filled and packed with the proper color of enamel it is leveled to the top of the wire.

Firing is now necessary. A temperature of 1450 to 1500 degrees F. is maintained for a period long enough to completely fuse the enamel.

Use of decorative gems to embellish a
pendant.

Swirling that is accomplished by using a
copper-tipped rod to manipulate molten
enamel lumps.

Daisy-type decorative gems used to embellish a tray.

Controlled design technique applied to molten enamel lumps.

The enamel tends to shrink slightly each time it is fired. When the work has cooled sufficiently it can be determined whether or not more filling and another firing is required. Sometimes a cloisonné looks beautiful and complete even though the wire is slightly higher than the enclosed enamel. If a level appearance is desired, more enamel and another firing will be needed.

The completed piece is pickled for a few minutes after which it may be lightly polished to bring out the luster of the metal wire.

The enterprising enamelist who wishes to extend his efforts will find many possible variations in the art of cloisonné.

Silver, gold-filled wire, or flat fine-silver ribbon may be used instead of copper wire.

Flat fine-silver ribbon is known as cloisonné wire and is often preferred to all other types of wire. It is normally eighteen-gauge wide and thirty-two-gauge thick. When it is placed upright on its narrow edge, deep recesses for the enamel are provided and fine lines and detail are achieved.

Finished piece of cloisonné.

13
Findings, Soldering, and Cementing

Included under the heading of findings are such items as pin backs, ear wires, jump rings, chains, and other miscellaneous accessories that are used in the construction of jewelry. These articles are usually manufactured and are not handcrafted. Much tedious work can be avoided by purchasing them ready-made.

Findings are obtainable in sterling silver, fourteen karat gold, and copper. They are also supplied in several plated finishes. Pin backs, for instance, can be had in silver-plate as well as in gold, rhodium, and nickle plate. Some findings such as cufflink backs and ear wires are provided purposely with large pads in order to be cemented readily. Others are manufactured with small pads and are designed especially to be soldered.

The technique used in soldering is quite simple, requiring two essential materials, soldering flux and solder. Soldering flux is a paste that can be obtained in small cans. Solder is composed of half tin and half lead and is purchased in the form of round wire. Since round wire tends to roll, it is wise to flatten it. This can be done with a hammer or by rolling it between jeweler's rollers. The thin ribbons of solder can be cut to various sizes and shapes with scissors.

Before a piece of jewelry is ready for soldering the fire-scale should be removed by immersing the piece in the pickling bath. The underside is then polished with steel wool until clean and free from any foreign material. A brush is used to apply a small amount of soldering flux to the space that is to be covered by the finding and also to the underside of the finding. Next, in sandwich fashion, the parts are assembled into one unit. The enameled piece, face down, is the bottom layer. A small ribbon of solder is the middle layer and the finding is the top layer. The

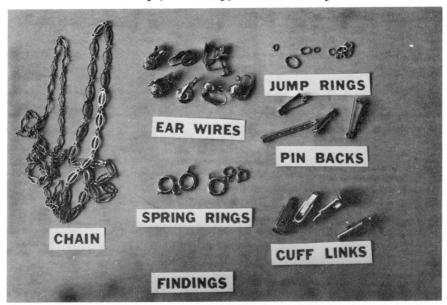

Typical findings.

combined unit then is heated gently with a torch, or placed on top of a hot kiln cover for a few seconds. Either method requires that the heat be applied to the enameled surface. If the torch is used, the article is placed enameled-side down on the metal grid and heated from the underside. Extreme care must be exercised since overheating could damage the finish or cause the enamel to stick to the grid. Heat eventually will melt the solder but will not damage the enamel if the torch is held at a distance and the flame turned low. Before long, the flux will begin to boil and the solder to melt and flow between the pieces. At this point, the article is removed and placed on an asbestos board to cool. The solder soon becomes rigid and the finding is held securely in place.

A cemented joint is not as strong as a soldered joint. Cementing, notwithstanding, is a very reliable method for attaching findings and is employed much of the time. If the piece has been counter enameled it must be cemented because solder will not adhere to enameled surfaces.

If the back of the article has not been counter enameled, it should be cleaned with steel wool before the finding is cemented to it. A small amount of epoxy cement is placed on the underside of the finding. The finding is then positioned, pressed down firmly, and set aside for several hours to harden.

Cleaning will be necessary if soldering has been done, but probably will not be necessary if the finding has been cemented. As a finishing touch the entire back can be lacquered.

14

Silver-Plated Steel

FACTS PERTAINING TO SILVER-PLATED-STEEL ENAMELING

(1) Precut shapes in silver-plated steel are available from suppliers. The price is comparable to that of copper shapes.

(2) Silver-plated steel is plated on both sides, the front side having a spangled pattern, which adds to the beauty of the finished piece.

(3) Transparent rather than opaque enamel should be used on silver-plated steel. The use of opaque enamel defeats the purpose because it obliterates the silver background.

(4) Transparent enamels are more brilliant when used over silver-plated steel than when used over copper. The silver shows through the enamel and intensifies the color.

(5) Silver-plated steel needs no cleaning after it has been fired. There is no accumulation of fire-scale and the work comes out of the kiln clean and bright.

(6) Counter enameling is not necessary except for the purpose of reinforcing the metal. The back as well as the front has a beautiful finish and will not tarnish.

(7) Steel, plated or not, is a poor conductor of heat. Due to this fact silver-plated steel requires special care in firing and should be fired in the kiln to assure a uniform temperature. If the torch is used, there is a likelihood that the piece may be over heated and the silver beneath the enamel melted.

(8) Silver-plated steel is an excellent base for most enameling and can be used in place of copper, except when the torch is required as in swirling or controlled design.

(9) Satisfactory crackled effects cannot be achieved over silver-plated steel.

(10) The finished enameled piece should be handled with care. The enamel does not adhere as tightly to silver-plated steel as it does to copper. Bending, dropping, or a sudden blow could cause the enamel to flake off.

METHODS USED IN ENAMELING SILVER-PLATED STEEL

(1) A thin coat of binder, either 7001 solution or klyr-fire, is applied to the surface.

(2) Number 450 clear flux is sprinkled over the binder. This flux is made especially for silver-plated steel. Flux that is used on copper is not satisfactory. It is not sufficiently transparent to derive the fullest benefit from the silver background.

(3) The work is fired in the kiln for approximately four minutes at about 1450°F. It is removed and allowed to cool.

(4) If a color coat is desirable, a binder and a coat of transparent enamel should be applied over the previously fired flux.

(5) After firing, the piece is ready for finishing, and from this point on the enameling procedure is similar to that used on copper. Sgraffito, stencil, overlay, and cloisonné are all used successfully on silver-plated steel. Lumps, threads, and other forms of ornamentation also can be used.

Sample shapes of silver-plated steel.

15
Thread Drawing

There is often a need for threads that are not commercially available. With the aid of a few tools and some special pieces of equipment the enamelist is now able to make threads of all sizes and colors.

Thread making requires the use of an acetylene torch, which should be equipped with a No. 3 tip, the one used most frequently by the beginner.

A drawing rod is also necessary and can be constructed from a ten-inch length of one-eighth inch stainless steel welding rod. In order to provide it with a handle, the rod is doubled back about an inch on one end. The other end is sharpened to a point by filing, grinding, or hammering.

A cup is needed in which to fuse the enamel. This should be made by the craftsman and should measure about one and one quarter inches in diameter. A piece of pipe may be used in the construction of the cup. The inside diameter of the pipe should be about one and one quarter inches and the length approximately one inch. A two-inch square of twenty or twenty-two-gauge copper is used to make the cup. The square of copper is placed on top of one end of the pipe. With a ballpein hammer, the central portion of the square is driven into the pipe, shaping the square into a cup. A block of wood provided with a hollowed-out, cuplike depression can be used in place of the pipe.

THREAD-DRAWING PROCEDURE

Approximately one-half teaspoonful of powdered enamel is put into the cup and the cup is placed on the grid. A slight depression or a hole in the grid provides a convenient place to set the cup. The heat from the torch is applied to the cup from the underside of the grid. When the enamel has become uniformly fused, the point of the drawing rod is

Typical tools needed for thread-drawing.

heated red hot and dipped into the center of the melted enamel. The rod is rotated a full turn in order to gather up enough enamel to hold it securely and prevent it from becoming disengaged. The rod is then raised upward drawing with it a thread of molten enamel. If the cup tends to be lifted up with the thread it is because the enamel is not as hot as it should be. When the thread has been drawn to the desired length it is separated from the remaining enamel by carefully guiding it to the rim of the cup and melting it free with the flame. The rod and the attached thread are then placed on an asbestos board to cool.

The thickness of the thread is influenced by the temperature of the enamel, the speed at which the thread is drawn, the type of enamel (hard or soft), and the size and shape of the drawing rod. A fine thread, for instance, will be obtained when soft enamel is used at high temperature and the thread is drawn rapidly. A coarse thread is produced when the enamel is barely melted and the thread is drawn slowly.

The cup requires cleaning before a new color is to be used. A satisfactory means of accomplishing this is to heat the cup red hot and plunge it into icewater. Most of the enamel will crack off and any that remains can be chipped off with a hammer. The tip of the rod can be cleaned in the same manner.

SPECIAL THREADS

Curved Threads

Curved threads can often be used to an advantage. They can be made by coiling molten enamel around the drawing rod. The rod is firmly embedded in the molten enamel and the thread is drawn. The rod is rotated or twirled in one direction, winding the molten enamel around

Box of various colored threads.

it and producing a continuous coil. The curvature of the coil increases in proportion to the increase in the distance between the cup and the rod. If small curves are desired, the rod should be twirled near the cup. If large curves are wanted, the twirling should be done at a greater distance from the cup. After the coil has cooled it may be cut or broken into a variety of sizes. The curves will be irregular but most of them can be used.

Bent Threads

Sometimes a bend in the thread is required. This is made by softening a small spot of the thread by holding it over a lighted birthday candle. The finer the thread the less heat will be needed to soften it. A right angle will be produced when the thread is held horizontally above the flame. When the thread becomes soft enough it will sag downward to a vertical position.

Beading

Beading is a term used to describe the balling-up or bunching effect that occurs when heat is applied directly to the end of the thread. The end of the thread is held above the flame of a match or small candle. The enamel fuses and crawls back on itself producing an accumulation of material. This beading technique is a valuable asset in certain types of ornamentations such as flower stamens and butterfly antennas.

Elongated Threads

Another technique can be employed by which cattails, teardrops, barley heads, petals, and other similar ornamentations can be made. A thread is held horizontally over a small candle flame. Tweezers are needed at this point to hold one end of the thread. The heat of the flame is concentrated on one spot of the thread. When the spot becomes soft and pliable the thread is removed from the flame. A slight pulling tension is quickly applied to it from either end, causing it to stretch and elongate into a fine thread. The various sections can be separated and used for special effects.

Straight Threads

Straight threads are cut to a suitable size and used for making stems, buds, petals, and numerous other ornamentations. The butterfly is an example of what can be accomplished with straight threads. The following paragraphs give specific directions for making a butterfly.

(1) A one-and-one-half-inch copper circle is enameled with a background coat of transparent enamel over clear flux.

(2) Four or six straight one-inch threads about one-thirty-second inch in diameter are needed for the butterfly wings. For best effects, more than one color of thread should be used. There should be two threads of each color, one for each side of the butterfly.

(3) These pieces are arranged on the enameled circle as illustrated and held in place with klyr-fire.

(4) The copper piece is then placed on the grid and heated with the torch. When the threads have become completely fused they are drawn with a copper-tipped swirling rod to form butterfly wings as shown in ·the second illustration.

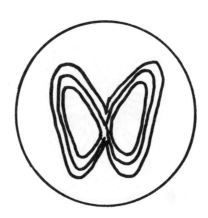

(5) After all four ends of each group of threads are drawn toward the center, the middle section of each group is also pulled to the center. This causes an excess amount of enamel at the junction, which must be removed. This is accomplished by heating a stainless steel rod and dipping it into the center of the accumulated enamel and drawing a thread. The thread can be disconnected by applying heat to the base of it.

(6) The antennas are made by using two short lengths of fine threads, which have been drawn and beaded according to previous directions. A half-inch section of slightly larger thread is used for the body. These threads are held in their proper position with klyr-fire or they may be placed on the work while it is hot and then fused.

(7) The work is refired and the final shaping of the wings is completed with a copper-tipped swirling rod.

The finished butterfly will be similar to the one in the following illustration.

16

Glass Beads

A recently developed process has made it possible for the hobbyist to make glass beads by using enamels. Pieces of red-hot copper tubing are rolled in enamel. The hot tubing picks up the enamel which is then fused.

NECESSARY MATERIALS

Copper tubing, which is one-eighth inch, three-sixteenth inch, or one-quarter inch, or larger in diameter, is used for the core of the bead. The tubing may be purchased in fifty-foot coils or in one-foot pieces and cut into suitable lengths for beads.

The following types of enamels are required:

(1) Soft fusing clear flux powder.
(2) White particles in eight-fourteenth mesh.
(3) Several selected colors of enamel in eighty-mesh.
(4) Several selected colors of threads and lumps.

NECESSARY EQUIPMENT

Two pieces of asbestos board, each approximately six inches by eight inches are needed. One board is used as a place for collecting and cooling the beads. The other one is used as a place upon which to put the various small piles of enamel that are used in making beads.

A torch is essential, and the acetylene torch is the ideal type for bead-making, but a propane or butane torch will suffice. The torch should be wired or clamped to the bench or grid as a safety measure.

A device is needed in order to cut the tubing into sizes suitable for beads. Information on this device is given in the following paragraphs:

(1) A block of wood approximately one-inch by four-inch by four-inch is securely fastened to the bench, the end-grained edge being uppermost. It can be fastened by using a vise, screws, or clamps.

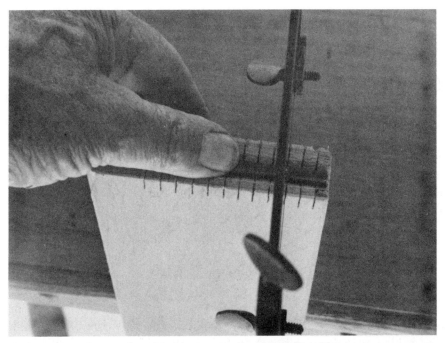

Sawing the tubing in preparation for bead-making.

Materials needed for bead-making.

(2) A "V" groove is sawed in the uppermost edge of the block. The groove should extend the full length of the block and be placed midway between the two sides. The depth of the groove should be no less than the diameter of the tubing.

(3) Saw cuts at one-quarter-inch intervals are made across the thickness of the block. They extend to the bottom of the groove and serve as a guide for measuring and cutting the tubing.

(4) The tubing is placed in the groove with one end flush with the end of the block.

(5) At a selected saw cut, the tubing is cut with a jeweler's saw using a five-inch "OO" blade. The tubing is then pushed over until it is again even with the edge of the block and another piece is cut in the same manner.

NECESSARY TOOLS

A firing rod is needed and it is made by using a nine-inch length of stainless steel welding rod. The rod should be slightly larger than the inside diameter of the tubing. About one-half inch of the rod is made smaller by filing to a diameter that allows it to fit into the tubing. A firing rod made by this method has a shoulder against which the tubing rest when it is being enameled.

A tool is needed to remove the hot finished bead from the firing rod. A knife or spatula can be used.

A six-inch mill file is needed to remove any exposed copper that may extend beyond the glass portion of the finished bead.

PROCEDURE FOR MAKING GLASS BEADS

(1) Several bead-length pieces of tubing are cut in the tube cutting device.

(2) A small amount of soft flux powder, a few grains of eight-fourteenth mesh white particles, and a small pile of the selected color of eighty-mesh powder are placed on one of the asbestos boards.

(3) The powders are flattened with a spatula to a thickness of about one-thirty-second of an inch.

(4) A piece of the tubing is placed against the shoulder on the firing rod.

(5) The torch is lighted and the tubing is heated until it is red hot.

(6) The hot tubing is rolled in the flux until the entire surface is coated.

(7) The tubing is again heated in order to melt the flux that has been gathered up on it.

(8) Now the hot tubing is rolled in eight-fourteenth mesh picking up a few white particles.

The tubing is again heated and must be continuously rotated while the particles are being fused, in order to maintain a balanced effect. At this stage, the bead begins to take on the final shape and size.

(9) Step eight is repeated as many times as necessary or until the bead is approximately the finished size and shape.

(10) Now the bead is reheated, rolled in the selected color of eighty-mesh powder and refired again. This operation must be repeated three or four times in order to produce a bead that is well coated and symetrical.

(11) As a final measure, the bead may be heated again, and threads or lumps added for special effects. The bead is now fired for the last time, rotating it continuously to ensure a uniformity of shape and temperature.

(12) The bead is allowed to cool slightly. It is then removed from the firing rod and placed on the asbestos board for further cooling. If permitted to remain on the rod for too long a period, it is likely to stick to the rod and be difficult to remove.

(13) Any exposed portion of the copper tubing should be filed away at this time. The bead is now completely finished and ready for stringing.

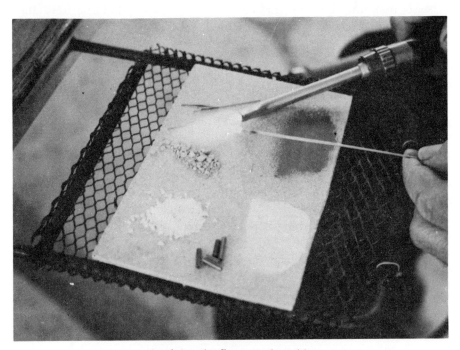

Applying the flame to the tubing.

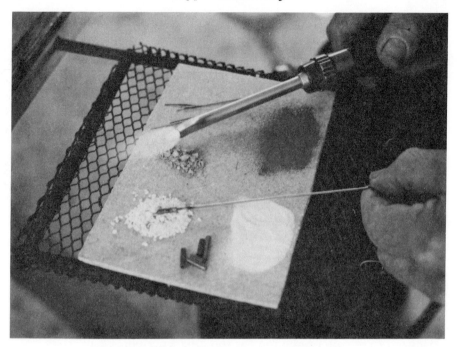

Picking up the lumps for buildup of the bead.

Removing the bead from the firing rod.

An assortment of handmade beads.

Various beads strung on leather lacing.

NOTES PERTAINING TO BEAD-MAKING

(1) The length of the bead is determined by the length of the tubing and the diameter of the bead by the amount of buildup.

(2) The shape of the bead is controlled by the angle at which it is held in the flame. Slanting the rod will cause the bead to be bulbous on one end. If the rod is held in a horizontal position a more uniform shape will result.

(3) To ensure uniform heating, the bead should be rotated continuously while it is in the flame. If one side becomes too hot the enamel will sag and the bead will become distorted. A bead that has been unevenly fired may have a tendency to crack or shatter.

(4) Sometimes the bead becomes discolored because of excess carbon in the flame. If this occurs, the bead should be moved to the outer portion of the flame where it is farther away from the torch tip. The outer portion of the flame burns hydrogen from the tank and oxygen from the air, so there is less chance for the bead to become carbonized or stained. Staining is a greater problem when transparent enamels are used. In most cases, the stain will disappear with additional heating.

(5) Transparent enamel powders used over opaque white produce especially pleasing effects.

INSTRUCTIONS FOR MAKING A BRACELET BY USING ENAMELED BEADS

(1) The core of the bead requires three-sixteenth-inch copper tubing.

(2) Twenty-two pieces of tubing are cut to three-quarter-inch in length.

(3) The pieces of tubing are coated with flux and three coats of selected enamel. Extra buildup is not needed, therefore eight-fourteenth mesh particles are not used. The beads are otherwise made according to the previously described procedure.

(4) When the twenty-two beads are finished, they are strung together using one and one-quarter yards of one-eighth-inch-round elastic. The elastic can be purchased at most notion stores and tinted to match the color of the beads.

(5) The first bead to be strung is placed midway between the two ends of the elastic. The second bead is strung parallel to the first and as close to it as possible. The two ends of the elastic are held firmly and passed through the second bead from opposite ends. The third bead is strung in the same manner and the procedure is continued until all of the beads are used. At all times, twisting of the elastic must be avoided.

(6) When the beads have all been strung, the two ends of the elastic are pushed through the first bead. The circle of beads is now complete and the elastic is tightened.

(7) The two ends of the elastic now protrude from opposite ends of the first bead. The pieces of elastic are stretched tightly and a knot is tied in each as close to the bead as possible. The excess elastic is cut off and the knots are tucked into the bead.

Beads and elastic used to make a bracelet.

Finished bracelets.

BEAD-MAKING TECHNIQUE AS APPLIED TO BOLO TIE TIPS

Preliminary Requirements

(1) A small piece of twenty-gauge sheet copper.
(2) A metal templet or paper pattern made according to the accompanying illustration.

PROCEDURE

(1) The templet or pattern is placed on the sheet of copper and traced with a sharp pencil or stylus.

(2) The copper shape is cut out with tin snips.

(3) The shape is formed into the core for the bolo tip by hammering it around a tapered piece of steel such as an awl. The two long edges should be joined without overlapping.

(4) The core is placed on the firing rod and enameled in accordance with the directions given for the bracelet bead, one coat of flux and three coats of a chosen powdered enamel being sufficient.

(5) As a final touch, the small end can be closed neatly by dipping it into the enamel powder and fusing it.

(6) The bolo tip is now finished and ready for removal from the firing rod.

Finished bolo tip before coating.

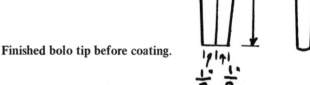

COMPLETING THE BOLO TIE

(1) A thirty-six-inch bolo cord, a decorative enameled tie slide, two bolo tips, and an appropriate finding are necessary. The cord and finding can be purchased and the enameled slide designed and made by the craftsman. The finding is soft-soldered to the back of the enameled slide.

17
Decorative Gems

Gem-making offers exciting possibilities and is a natural followup to thread-drawing.

Small wafers of enamel that are made especially to be used as embellishments are known as gems. They are placed on prefired enameled surfaces or on unfired surfaces that have been dusted with enamel powder. The undercoat and the gems are then fired in one operation by using either the kiln or the torch.

ESSENTIAL TECHNIQUES

Gems are made by arranging enamel powder, lumps, and threads in a copper cup like the one described in the previous chapter on thread-drawing. The enamel in the cup is fused by using a torch, and a thread is drawn according to the instructions given in the chapter on thread-drawing.

After the thread is cool it is cut into small sections, which are the finished gems, the thickness of these gems being about one third of their diameter.

COMPOSITION

It is necessary to understand what actually happens when two or more colors are simultaneously fused in the same cup and a thread is drawn. The color that is placed in the bottom of the cup becomes the center of the finished gem. The next color to be placed in the cup surrounds the first color. Any additional colors, according to their position in the sequence, will be closer and closer to the outside rim of the gem. If the gem, for instance, is to be red with a white center, the white will be used to fill the bottom half, and the red, the remaining half. The thread produced from this combination will be red but will have a white center running parallel to its length. Gems cut from this thread will be, as a rule, red with white centers. Should the cup be filled with half blue and half green, each color occupying one side of the cup, the thread will have a green side and a blue side running the full length of it, and the gems will be half green and half blue.

GEM CUTTING

The gems are cut with tile-worker nippers. Here a problem presents itself because the gems have a tendency to scatter in all directions when being cut. This problem is solved by employing either of the following measures.

Transparent Bag Method

A small transparent plastic bag is required. One end of the thread is pushed into the bag through a small hole in the closed end. The thread remaining outside of the bag is held firmly with one hand. The other hand is free to manipulate the nippers. The nippers are held inside the bag where the gems are cut and accumulated. It is possible to see through the plastic to determine whether or not the gems are being cut properly.

Nippers with Added Aprons

This is the sophisticated method and it should be resorted to when a considerable amount of cutting is necessary. An improvised apron or guard is attached to each side of the nippers to prevent the particles from escaping. These aprons are about one inch in diameter, made of copper and shaped to fit the jaws of the nippers. They are fastened to the nippers with small bolts or machine screws.

Cutting gems with nippers.

METHODS USED IN MAKING SPECIAL GEMS

The most outstanding gems are those that portray flowers. In the following paragraphs, instructions are given for making a variety of flower gems. All of these are made by using medium-fusing enamel. The student should not confine his efforts to these specific examples. He will discover by experimentation that he can create many beautiful and original gems.

A Fundamental Type (Sunflower used on any background)

(1) About one-half teaspoonful of opaque brown enamel is placed in the bottom of the cup and leveled and packed with the spoon.

(2) The cup is then filled to within one-eighth inch of the top with opaque yellow enamel and is leveled and packed.

(3) The next step requires the use of threads, these having been previously made of the same brown that is used in the bottom of the cup. There should be twelve to sixteen pieces of thread, each approximately one-thirty-second inch in diameter and three-eighth inch in length.

With the aid of tweezers, the threads are placed on the yellow enamel. They are arranged in a radiating pattern, touching each other near the center of the cup and pointing outward toward the rim. These threads become the petals of the finished flower.

(4) The enamel is now ready to be fused and the thread to be drawn.

Making the fundamental type gem with threads and powder.

Conventional White Flower (To be used on a blue background)

(1) The cup is filled and packed with opaque blue enamel to within one-eighth inch of the top.

(2) A hole or well is made in the center by pushing the unsharpened end of a pencil through the enamel to the bottom of the cup. This is done cautiously in order to prevent the enamel from caving back into the hole.

(3) The hole is then filled with opaque enamel. This enamel will be the center of the flower. In order to fill the hole easily a small funnel is needed.

(4) Five to eight white lumps about the size of small peas are selected. They are placed on the blue enamel in a uniform pattern midway between the center and the rim.

(5) The enamel is now ready for fusing and thread-drawing.

In order to obtain the most desirable effect, the finished gems are used on a surface that is enameled with the same blue enamel that is used in step one. The blue in the gem is identical to the blue in the background and will not show in the finished gem.

Conventional White Flower (To be used on contrasting background)

When these gems are fired on any color other than the opaque blue the effect will be quite different. The opaque blue in the gem will appear as a matrix surrounding the petals and the center. The flower may not be as outstanding as the one fired on the opaque blue but it will be very satisfactory. By using different backgrounds any number of effects can be obtained.

A Simple White Daisy Type

(1) The cup is packed to the top with opaque chinese red enamel.

(2) Eight white threads, one-thirty-second inch in diameter and three-eighth inch in length, are placed evenly on the red enamel, touching at the center and radiating to the rim of the cup.

(3) The cup of enamel is fired and the gem threads are drawn.

These gems are easy to make and are very dependable. They are at their best when used on a Chinese red background, but can be used successfully on most any background.

Conventional method of making gems.

The Cutting-In System (using powdered enamel)

(1) The cup is filled to within one-eighth inch of the top with three different colors of opaque enamel; yellow, white, and orange. White occupies one-quarter of the space, yellow occupies two fourths, one fourth adjoining each side of the white, and the remaining one-quarter is filled in with orange. The two yellows are opposite and the white and orange are opposite.

(2) The enamel is firmly packed and then six grooves or wedges are made in it with the palette knife. The grooves extend to the bottom of the cup and are about one-eighth inch wide at the rim tapering off as they reach the center.

(3) The grooves are filled with medium-fusing flux. They must be made and filled one at a time in order to prevent the walls from caving in. With the aid of the palette knife, each groove is made and filled with flux before proceeding to the next groove.

(4) Upon completion of the last groove the enamel is firmed. A hole is made in the center of the enamel and filled with light brown enamel.

(5) The enamel is then fired and the gem thread is drawn. These gems can be used on any contrasting background: transparent green is suggested.

The cutting in method.

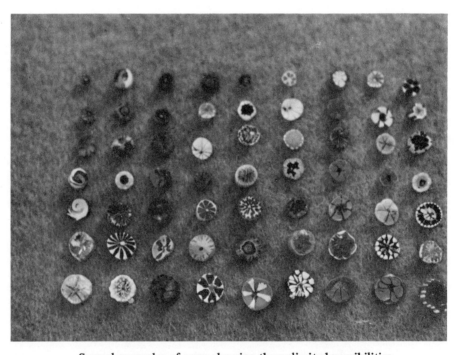

Several examples of gems showing the unlimited possibilities.

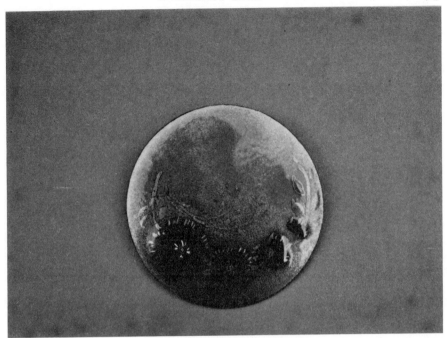

Another example of the use of gems.

Gems Used As Petals of Flowers

Gems can be made especially to be used as single flower petals. They are a delightful variation of the usual gem and are very easy to make.

(1) The cup is filled to within one-eighth inch of the top with any chosen color of enamel, and the enamel is packed down and leveled.

(2) With the palette knife a single groove or wedge is made in the enamel from the rim of the cup to the center. The wedge extends to the bottom of the cup and is about one-eighth inch wide at the rim, the sides of the wedge coming together in the center.

(3) The wedge is filled with either clear flux, opaque, or transparent enamel.

(4) The enamel is fused, a thread is drawn, and the gems are cut.

Gems made by this method are used as individual petals and five or more are assembled to make a flower. They are arranged as petals, the large ends of the wedges being placed near the center of the flower.

(5) A lump or cluster of small fragments of some suitable color can be used as the center of the flower.

Gems Used As Leaves

Gems can be made especially to represent leaves. The method used in making leaves is the same as that used in making petals. The flux or powdered enamel that is placed in the wedges becomes the veins of the leaves.

Applying the Leaves

(1) The leaves are positioned on the enameled piece with the wedge sides as the bases of the leaves.

(2) The piece is heated with the torch until the enamel is completely fused.

(3) Using the copper-tipped swirling rod, the ends of the leaves are pulled out slightly to make them appear pointed.

(4) The bases of the leaves are pulled inward to give them a more natural leaflike appearance.

Creating with decorative gems.

Typical use of decorative gems.

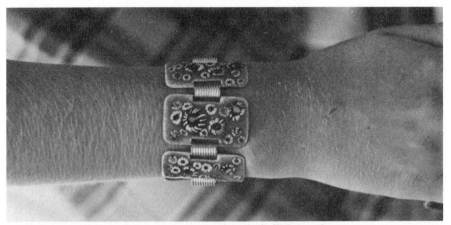

Decorative gems used to embellish jewelry

Decorative gems used to embellish jewelry.

Index

Acetylene Gas, 22
Acetylene Torch, 38, 70, 76
Acid (Nitric), 16
Acid (Sulphuric), 16
Agar Solution, 17
Aluminum, 13, 14
Asbestos Board, 34, 43, 67, 71, 76, 79

Bamboo Stick, 47, 64
Base Coat, 64
Base Metal, 13, 14, 29
Beading, 73
Beads, 76, 78, 79, 82
Bent Threads, 73
Binders, 16, 31, 38, 47, 69
Bleeding, 30
Blisters, 14, 31, 43
Bolo Tie Tips, 84
Bracelet, 82
Brass, 14
Bubbles, 14
Butane Torch, 76

Candle Flame, 73
Cementing, 66, 67
Clear Flux, 69, 74, 76, 91, 92
Cloisonné, 64, 65, 69
Controlled Crackle, 43
Controlled Design, 49, 68
Copper, 13, 14, 16, 28, 29, 31, 55, 56, 66, 69, 70, 86
Copper Cup, 70, 85, 87, 88, 89, 91
Copper Disc, 49
Copper Piece, 37, 38, 64
Copper Shapes, 15, 31, 68, 74, 84
Copper Sheet, 13, 84
Copper Spatula, 18, 59
Copper-tipped Rod, 18, 49, 74, 78, 92
Copper Tubing, 76
Copper Wire, 64, 65

Counter Enameling, 35, 37, 42, 43, 48, 53, 67, 68
Crackling, 29, 39, 42, 59
Curved Threads, 72
Cutting Device, 76

Drawing Rod, 70, 71, 72

Electric Kiln, 15
Element, 19
Enamel, 14, 15, 16, 19, 26, 28, 29, 30
Enamel Characteristics, 28
Enamel (Crackle), 29, 42
Enamel (Hard or High-firing), 29
Enameling Kits, 14, 28
Enamel (Low-firing or Soft), 14, 29
Enamel (Medium-firing), 29
Epoxy Cement, 67
Expansion of Copper, 35

Findings, 37, 66, 67, 84
Finishing, 43, 47, 69
Firescale, 16, 34, 39, 43, 66, 68
Firing, 26, 27, 43, 47, 48, 55, 57, 64, 65
Firing Fork, 22
Firing in Kiln, 31
Firing Rod, 78, 79, 84
Firing Temperature, 29
Firing with a Torch, 31
Fluxes, 28, 29, 30, 61, 62, 66, 67, 69, 76, 78, 79, 82, 89, 91, 92
Foil (Copper), 60
Foil (Gold), 60
Foil Overlay, 60, 61, 62
Foil (Silver), 60
Frit, 28

Gems, 85, 86, 87, 88, 91, 92
Glass Beads, 76

Glass Sheets, 29, 53, 64
Gold, 13, 14, 62, 66
Gold-filled Wire, 65
Gold Overlay, 62
Gold Plate, 66
Grid, 26, 27, 31, 38, 39, 59, 67, 70, 76
Gum Tragacanth, 16, 31, 47

Hearth, 19, 21
Heat Lamp, 16, 42, 47

Identification System, 14

Kilns, 18, 19, 22, 37, 43, 53, 57, 59, 60,
   68, 69, 85
Klyr-Fire, 16, 17, 19, 47, 64, 69, 74

Low-firing Enamels, 30, 58
Lumps, 28, 38, 49, 58, 69, 79, 88, 91

Masking, 53, 54, 55, 56
Metal (Base), 13, 14, 29
Metallic Oxides, 28

Nippers, 18, 86
Nitric Acid, 16

Opaque Enamels, 47, 61, 64, 68, 87, 88,
   91
Open Stencil, 53
Overfiring, 29, 30, 58
Overlay, 61, 69

Palette Knife, 18, 29, 43, 57, 64, 89, 91
Pickling Solution, 16, 30, 34, 39, 43, 66
Precut Shapes, 13, 68
Propane Torch, 22, 76
Protectors, 16

Refiring, 30
Rhodium Plate, 58

Scalex, 16, 17
Scissors, 18, 53, 60, 66
Sgraffito, 47, 48, 69
Shapes (Precut), 13, 68

Silver, 13, 14, 29, 62, 68
Silver Foil, 60
Silver Overlay, 62
Silver Plate, 66
Silver-plated Steel, 13, 14, 29, 68, 69
Silver Ribbon, 65
Silver Wire, 65
Simple Crackle, 42, 43
Simple Masking, 54, 55
Slush, 29
Solder, 17, 60, 67, 84
Soldering Fluxes, 17, 60
Solution (Agar), 17
Solution "7001," 15, 17, 31, 37, 38,
   47, 49, 53, 54, 55, 56, 61, 69
Sparex No. 2, 16, 17
Spatula, 15, 31, 39, 60, 78
Special Gems, 87
Special Threads, 72
Special Torch, 26
Stainless Steel, 21
Stainless-steel Rod, 18, 49, 58, 59, 75
Stainless-steel Spatula, 18
Starter Set, 14
Steel Rod, 60
Steel Wool, 31, 38, 60, 66, 67
Sterling Silver, 66
Stenciling, 53, 69
Stilts, 21
Straight Threads, 73
Stylus, 47
Sulphuric Acid, 16
Swirling, 15, 38, 39, 68

Threads, 38, 49, 57, 69, 70, 72, 73, 74,
   75, 79, 85, 86, 87, 88
Tooling Copper, 60
Torches, 22, 38, 39, 48, 49, 53, 58, 60,
   67, 68, 70, 74, 76, 78, 85
Tracing Paper, 53, 54, 61
Trivets, 21, 37
Tubing (Copper), 78, 79
Tweezers, 18, 60, 61, 62, 70

Undercoat, 47, 48, 58, 61